Just Hanging Around

MARTHA CIPOLLA

SCHOLASTIC INC.

Copyright © 2019 by Scholastic Inc.

All rights reserved. Published by Scholastic Inc., *Publishers since 1920.* SCHOLASTIC and associated logos are trademarks and/or registered trademarks of Scholastic Inc.

The publisher does not have any control over and does not assume any responsibility for author or third-party websites or their content.

No part of this publication may be reproduced, stored in a retrieval system, or transmitted in any form or by any means, electronic, mechanical, photocopying, recording, or otherwise, without written permission of the publisher. For information regarding permission, write to Scholastic Inc., Attention: Permissions Department, 557 Broadway, New York, NY 10012.

ISBN 978-1-338-57278-0

10 9 8 7 6 5 4 3   20 21 22 23

Printed in the USA   40

First printing 2019

Illustrations by Michelle Simpson

Book design by Erin McMahon

WELCOME TO THE ANIMAL KINGDOM!

GRAB YOUR FRIENDS AND DIVE INTO A MENAGERIE OF COOL QUIZZES, ACTIVITIES, JOKES, AND MORE. BECAUSE WHETHER YOU PREFER THE DESERT, THE OCEAN, OR SOMETHING IN BETWEEN, IT'S TIME TO GET IN TOUCH WITH YOUR *wild side!*

SPOTS or STRIPES?

EVERY ANIMAL HAS ITS OWN FAB FASHION—WHETHER STRIPED, SPOTTED, OR PAISLEY—AND SO DO YOU! TAKE THIS QUIZ TO FIND OUT WHICH ANIMAL INSPIRES THE WAY YOU STRUT YOUR STYLE.

1. YOU GET TO TAKE A DREAM TRIP WITH YOUR BESTIES. WHERE DO YOU WANT TO GO?
    a. An outdoor adventure camp. Hello, zip lines!
    b. Hollywood for photo ops and touring the set of your favorite TV show.
    c. Chill days floating in the neighborhood pool, plus an endless supply of Popsicles.
    d. Paris. Croissants, window-shopping, and world-class art? *Très chic!*

2. WHICH ACCESSORY ARE YOU MOST LIKELY TO WEAR?
    a. A headband.
    b. A huge, sparkly pendant.
    c. A pair of delicate gold studs.
    d. A heart-shaped ring.

3. WHAT'S THE BEST WAY TO SPEND SATURDAY WITH YOUR FRIENDS?
   a. Rollerblading in the park.
   b. Choreographing a dance together.
   c. An all-day board-game tournament.
   d. Crafting gorgeous paper flowers.

4. WHAT IS YOUR FAVORITE DESSERT?
   a. A slice of pie.
   b. A sundae with literally all of the toppings.
   c. A perfect cupcake.
   d. Chocolate-dipped strawberries.

5. WHICH MOVIE SOUNDS THE BEST TO YOU?
   a. Something with lots of adventure.
   b. One about a kid who becomes a movie star.
   c. A funny movie about a talking cat.
   d. One about magical elves.

6. WHAT KIND OF SUMMER JOB WOULD YOU LOVE?
   a. Something outdoorsy, like being a junior camp counselor.
   b. Run a stand selling lemonade and homemade art.
   c. No job at all—that would get in the way of your reading time!
   d. Babysitting, dog walking, mowing lawns—anything to help you fill up that piggy bank.

7. HOW DO YOU LIKE TO SHOP FOR CLOTHES?
   a. Ugh, shopping!
   b. Check out thrift stores to find a one-of-a-kind look that's as unique as you are.
   c. Window-shop till you drop, then go back and snag all of your faves.
   d. Pick out the hottest looks from your go-to magazines, then hit the mall to put together your own version.

If you got mostly As, you're a *sporty zebra*! You love to stay active in clothing that is comfortable and practical. You can give your athletic look zebra-esque pizzazz by choosing fabrics in colors and patterns that you love. (Distinctive stripes are always a strong choice!) And don't be afraid to glam it up occasionally with a chunky bracelet or snappy pair of shoes.

If you got mostly Bs, you're a *flashy peacock*! You express your creative side through clothing by showing off your funky feathers. Look for bold patterns, bright colors, and big costume jewelry, and mix and match things that aren't usually paired together. Do snap up some solid-colored tops and bottoms, though, to go with extra-busy pieces—and to give yourself the option to let your feathers down sometimes.

If you got mostly Cs, you're a *classic deer*! You love to look cute and wear current fashions, but without drawing too much extra attention. Deer blend in and stand out with a combo of neutral colors and spotted coats, and you can, too. Solid colors, simple stripes, small polka dots, and understated gold and silver jewelry are always in style. But consider adding a sparkly cardigan, bright-colored shoes, or some stand-out accessories to your closet for the occasional deer jamboree.

If you got mostly Ds, you're a *romantic unicorn*! Gauzy, lacy whites and pastels will give your style a dreamy vibe, and carefully placed rhinestones and pearls are your friends. You can use ribbons to add a mythical touch to your outfits by tying one around your wrist as a bracelet, tying one to your belt loop, or weaving one into your hair with a loose braid. Unicorns do like to frolic in the woods, though, so keep your closet balanced with a few pairs of blue jeans and some dark-colored tops.

CHINCHILLIN'

WHAT'S SUPER SOFT, SUPER SWEET, AND NEEDS SUPER SPECIAL CARE?

Chinchillas, and also friendships!

TO CARE FOR CHINCHILLAS, YOU LET THEM TAKE A DUST BATH. AND TO CARE FOR FRIENDSHIPS, YOU

1) ARE CONSIDERATE OF YOUR FRIENDS AND

2) HAVE A GREAT TIME CHINCHILLIN' WITH THOSE FRIENDS.

Llama Drama

Llamas are very social and friendly, and they love hanging out with their buds. But they can also be pretty dramatic—when they get in fights, they even spit at each other! The good thing about llamas having a spit-fight is that it allows them to get all their feelings out in the open and resolve the conflict. The bad thing is all the spit.

There are a lot of things you and your friends can do to keep the llama drama in your own friendships to a minimum. Check out these suggestions together!

The main thing is to always practice the golden rule: Treat others the same way you would like to be treated. Here are some specific ways you can do that:

- Be honest with your friends. Tell them the truth, don't talk about them behind their backs, and (politely) share your real feelings with them.

- Make sure your friends can trust you. Don't tell other people their secrets. (An exception is if they—or someone else—are being hurt by the secret. Then you should tell an adult you trust.)

- Be a good listener. Make sure to give your friends a turn to talk. When they do, pay attention to what they're saying. You'll learn more about them that way!

- Do thoughtful things for your friends. Get your friend their fave candy bar just because. If one of your friends is feeling down, draw them a picture or write them a poem. And always remember your friends' birthdays so you can help them celebrate!

- Be considerate of your friends' needs. If you invite friends who have food allergies over, be sure there's food they can eat. If you have a friend who's afraid of lightning, text them to check in during thunderstorms.

- Talk to your friends if something they're doing is hurting your feelings. This can seem scary at first, but if you talk to them about it calmly, it will make you better friends in the long run. And if a friend tells you something *you're* doing is hurting *them*, hear them out! Either way, talk together about what the two of you can do to make sure you're both happy.

Llama Bonding Time

SOME THINGS YOU CAN DO TO MAKE YOUR FRIENDSHIPS EVEN BETTER ARE: CREATING A SCRAPBOOK TOGETHER, MAKING MATCHING BEADED BRACELETS, WRITING EACH OTHER LETTERS ABOUT THINGS YOU LIKE ABOUT EACH OTHER, AND DOING A "SKILLS EXCHANGE" WHERE YOU TEACH EACH OTHER SOMETHING NEW.

A FLAMBOYANCE OF *Flamingos*

You've heard of a herd of antelope and a gaggle of geese, but what about a flamboyance of flamingos or a kindle of kittens? Scientists don't use these words to describe groups of animals, but they *are* real terms—they originate all the way back to the Middle Ages.

- WADDLE OF PENGUINS
- MURDER OF CROWS
- BED OF SLOTHS
- PEEP OF CHICKENS
- SLOTH OF BEARS
- BUSINESS OF FERRETS
- OSTENTATION OF PEACOCKS
- SIEGE OF HERONS
- PARLIAMENT OF OWLS
- CRASH OF RHINOCEROSES
- RAFTER OF TURKEYS
- UNKINDNESS OF RAVENS
- SKULK OF FOXES

What are some group names you could use for you and your besties? What about your family, your sports team, or your class at school? What could you call a group of musicians, teachers, doctors, astronauts, or farmers? Let your imagination run wild!

Sneaky
LIKE A FOX

> FOXES ARE MASTER SPIES, RENOWNED FOR THEIR FLUFFY TAILS AND THEIR SNEAKY WAYS. OH, YOU HAVE AN EGG? NO, YOU DON'T. YOU'RE HOLDING A ROCK AND THAT FOX OVER THERE JUST MADE AN OMELET!

But don't despair. You, too, can be super sneaky! Sneaky like a fox. A fox who writes messages in secret code! Create your very own code with your friends so you can be master spies and write messages about all of your daring adventures with no one else being any the wiser.

To get started on your code, write out the alphabet on a sheet of paper, one letter per line. Then work with your friends to pick an animal that starts with each letter. To write notes in code, use the full animal name instead of the letter, and don't use any capital letters or punctuation. This way, if someone finds your note, they won't know what it says. They will, however, be confused by your list of random animals. And their confusion is your success!

HERE'S AN EXAMPLE: YOU WANT TO END YOUR MESSAGE WITH "BFFS," AND YOUR CODE INCLUDES THESE WORDS:

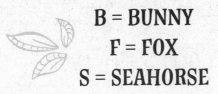

B = BUNNY
F = FOX
S = SEAHORSE

SO YOU WRITE:
BUNNY FOX FOX SEAHORSE

Animal Emporium

SOME LETTERS MIGHT BE EASIER
TO FILL IN THAN OTHERS!
IF YOU GET STUCK, TRY THESE:

N = NARWHAL, NEWT, NIGHTINGALE
Q = QUAIL, QUOKKA
U = UNICORN FISH, URCHIN, URIAL
V = VULTURE, VIPER, VOLE
X = XERUS
Y = YABBY, YAK

PET PSYCHIC

PREDICTING THE FUTURE ISN'T JUST FOR PEOPLE WITH MYSTERIOUS TENTS AND CRYSTAL BALLS.

It's also for pets!

FIND OUT WHAT YOURS HAS TO SAY ABOUT WHAT'S IN STORE FOR YOU AND YOUR FRIENDS.

1. Sit in a circle. Each player writes three predictions on separate slips of paper for the player to their left.
2. For each player's turn, place the three predictions facedown in a row on the floor and place one of your pet's favorite treats on top of each one.
3. Use another treat to lure your pet to the fortune-telling room. (Only play this game with willing pets!)
4. Sit the pet down a few feet away from the predictions. Whichever one your animal pal goes for first will come true!

If your pets aren't in a fortune-telling mood, nix the treats, spread the predictions a little farther apart, and hold a small stuffed animal over them. Close your eyes, move your arm back and forth, then release the stuffed animal.

Whichever slip of paper it lands closest to is its prediction!

WHALE of a TAIL

Oh, hey, it's story time!

PUT YOUR HEADS TOGETHER TO TELL AN AWESOME ADVENTURE TOGETHER. BUT DON'T PEEK AT THE STORY FIRST! ONE PERSON WILL READ OUT WHAT KINDS OF WORDS ARE MISSING, AND EVERYKITTY ELSE WILL TAKE TURNS SUGGESTING WORDS TO FILL IN THE BLANKS. THEN THE WORD COLLECTOR CAN READ THE STORY OUT LOUD!

Grammar Refresher

A **NOUN** IS A PERSON, PLACE, OR THING. • A **VERB** IS AN ACTION. • AN **ADJECTIVE** DESCRIBES A NOUN. • AN **ADVERB** DESCRIBES AN ADJECTIVE OR A VERB AND OFTEN ENDS IN "LY."

Sammy the _____ (animal) and his best friend, Gruff the _____ (different animal), were dancing along the seashore during a _____ (adjective) nighttime beach bash when Sammy gasped, "Look! there's a _____ (noun). Someone must have lost it!"

"We should keep it!" Gruff said _____ (adverb). "Finder's keepers! You snooze, you _____ (verb)! The early bird gets the _____ (noun)!"

"Well," Sammy said, "it is very _____ (adjective). I'd like to keep it, too. But first, let's ask if it belongs to anyone else at the party."

They asked a _____ _____,
                    (adjective)   (animal)
and a _____-playing _____.
          (sport)   (animal)
They even asked a very tall _____. But it
                                  (noun)
did not belong to any of them.

All the asking made Gruff and Sammy hungry for some _____. While they were eating,
                                (noun)
a _____ waving a glow-in-the-dark
     (animal)
_____ cartwheeled up and said, "Hi, I'm
    (noun)
Maria!" Then she pointed at the object, exclaiming, "Oh, good, you got my present!"

Maria smiled. "I left that by the water for someone at the party to find. Isn't it _____?"
                                                       (adjective)
"That was so nice of you! Thank you!" Sammy said. Then he held his plate out to Maria. "Want some?"

Then Maria, Sammy, and Gruff sat down together and watched as the _____ started to
                                  (plural animal)
dance under the sparkling stars. It was the beginning of a _____ friendship.
                (adjective)

WHICH PAIR OF ANIMAL BEST FRIENDS ARE YOU?

LISTEN. ANIMALS OF DIFFERENT SPECIES WHO ARE BEST FRIENDS ARE THE ACTUAL CUTEST THING IN THE WORLD. AND YOU KNOW WHAT ELSE IS THE ACTUAL CUTEST THING IN THE WORLD? YOU AND YOUR OWN BEST FRIEND!

TAKE THIS QUIZ TO FIND OUT WHAT PAIR OF ANIMAL PALS YOU HAVE THE MOST IN COMMON WITH.

1. WHEN YOU AND YOUR BESTIE HAVE SLEEPOVERS, WHERE DO YOU CRASH?
    a. Snuggled up in one bed.
    b. Camped out on the living room floor.
    c. In a blanket fort.
    d. You flip a coin for the bed; the other one sleeps on the floor.

2. THE TWO OF YOU ARE HAVING SOME POPCORN. HOW DO YOU EAT IT?
   a. Share it in a big bowl.
   b. Split it evenly into individual bowls.
   c. Throw it for each other to catch in your mouths.
   d. Take turns adding different spices to a few pieces at a time to come up with your own popcorn flavor.

3. WHEN IT COMES TO ACCESSORIES, YOU AND YOUR BESTIE . . .
   a. Share everything, from socks to tiaras.
   b. Stick to your own closets.
   c. Pick out silly things for each other to wear.
   d. Get each other to try out new styles.

4. YOU AND YOUR PAL ARE AT THE MALL. WHAT DO YOU DO?
   a. Link arms and check out all the novelty shops.
   b. Head to a department store and split a dressing room so you can yea or nay all of each other's threads.
   c. Hide behind various plants to spy on unsuspecting shoppers.
   d. Dare each other to talk to cute cashiers at the food court.

5. YOUR BEST FRIEND CALLS RIGHT BEFORE PHONE CURFEW. WHAT DO THEY SAY?
   a. "Just calling to say good night. Sweet dreams!"
   b. "Can I borrow your sequined cardigan tomorrow?"
   c. "[unintelligible animal noises]"
   d. "I have a crazy idea. Are you in?"

6. WHEN YOU'RE JUST HANGING OUT, YOU'RE MOST LIKELY TO BE . . .
   a. Talking up a storm.
   b. Playing basketball.
   c. Goofing off.
   d. Acting out the best parts of your favorite movies.

7. YOU DECIDE TO HAVE A JOINT STUDY SESSION. WHAT DO YOU DO?
   a. Quiz each other out loud.
   b. Study individually, but ask each other for help if you have questions.
   c. Turn it into a game to see who knows the material best.
   d. Accidentally get distracted playing paper football.

If you got mostly As, you and your bestie are a *kitten and a duckling*! You love to snuggle, chat, and just hang out together—but you're also up for some goofing around, when the mood strikes.

If you got mostly Bs, you and your bestie are a *deer and a horse*! You're an equal match for each other and are great at compromising, which makes it easy to have a top-notch time whenever you're together.

If you got mostly Cs, you and your bestie are a *parakeet and a puppy*! Your top priority is maximizing the silliness factor in any situation, though you can also settle down and be serious if you absolutely have to.

If you got mostly Ds, you and your bestie are a *bear and a tiger*! You take turns being the bear, who instigates exciting adventures (that, ahem, may not always be totally permitted). Just be careful not to get yourselves into real trouble!

Purrsonal CARE

SOME PURRFECT PURRSONAL CARE IS A TERRIFIC TREAT TO CELEBRATE NAILING A BIG PERFORMANCE, REACHING THE END OF A SPORTS SEASON, ACING A TEST, OR JUST THE FACT THAT IT'S TUESDAY.

Cheshire Cat
LIP SCRUB

WHEN YOU'RE GRINNING LIKE A CHESHIRE CAT, YOU WANT TO MAKE SURE THAT GRIN LOOKS GOOD! GENTLY EXFOLIATING YOUR LIPS KEEPS THEM SMOOTH AND SOFT AND HELPS AVOID CHAPPING.

Supplies

1 TEASPOON HONEY

2 TEASPOONS SUGAR

LIP BALM

SUGAR IS A GREAT NATURAL EXFOLIANT, WHICH MEANS IT HELPS REMOVE DEAD SKIN.

Instructions

1. MIX THE HONEY AND SUGAR TOGETHER IN A SMALL BOWL.

2. GENTLY MASSAGE THE MIXTURE INTO YOUR LIPS.

3. LEAVE IT FOR 10 MINUTES.

4. WIPE IT AWAY WITH A SOFT CLOTH DIPPED IN WARM WATER.

5. USE YOUR FAVORITE MOISTURIZING LIP BALM.

AVOCATO FACE MASK

CATS ARE ALWAYS READY TO STOP EVERYTHING AND LAUNCH INTO A TONGUE BATH, WHICH KEEPS THEM CLEAN AND HEALTHY. GIVE YOUR SKIN A HEALTHY BOOST OF ITS OWN WITH A GOOEY GREEN AVOCADO FACE MASK.

Supplies (per person)

- HALF AN AVOCADO
- 1 TEASPOON PLAIN YOGURT
- 1 TEASPOON HONEY

THE OILS IN THE AVOCADO ARE HYDRATING. THE LACTIC ACID IN THE YOGURT IS AN EXFOLIANT. AND THE HONEY HAS ANTIBACTERIAL PROPERTIES.

Instructions

1. SCOOP THE AVOCADO INTO A BOWL AND MASH IT WITH A FORK UNTIL IT'S SMOOTH AND EASY TO SWIRL AROUND.

2. ADD IN THE YOGURT AND HONEY AND STIR UNTIL THE MIXTURE HAS A PASTE-LIKE TEXTURE.

3. SPREAD IT ON YOUR FACE. LEAVE A BARE CIRCLE AROUND EACH EYE, FROM JUST ABOVE THE EYEBROW TO JUST BELOW THE LOWER EYE SOCKET BONE.

4. LEAVE IT ON FOR 15 MINUTES.

5. WASH IT OFF WITH LUKEWARM WATER. YOUR SKIN WILL FEEL SMOOTH AND REFRESHED!

Claw Care
MANICURES

YOU HAVE MORE IN COMMON WITH CATS THAN YOU MIGHT THINK. BOTH YOUR FINGERNAILS AND THEIR CLAWS ARE MADE OF THE PROTEIN KERATIN, AND BOTH ARE ALWAYS GROWING. SO YOU WANT TO KEEP THEM STRONG AND CLEAN. HERE ARE SOME DOS AND DON'TS:

DON'T trim your cuticles or push them back. They protect your fingernails from infections and should be left alone.

DO use hand lotion every day. This helps keep your nails from cracking, and it helps your cuticles stay nice-looking, too.

DON'T change nail polish too often. Nail polish remover is tough on your nails.

WHEN THE COWS COME HOME

SHOW OFF YOUR MOO-VELOUS FASHION SENSE WITH COW PATCH FINGERNAILS. YOU'LL NEED BLACK AND WHITE NAIL POLISHES AND A TOOTHPICK FOR THIS PATTERN.

1. Paint your nails white and allow to dry completely.
2. Set the black nail polish brush aside. With a toothpick, draw one or two patches on each nail. Allow to dry completely.

MER-MADE FOR FASHION

SCALE UP YOUR STYLE WITH SLEEK MERMAID FINGERNAILS. FOR THIS PATTERN, YOU'LL NEED A SHIMMERY POLISH IN BLUE OR GREEN, WHITE POLISH, AND A TOOTHPICK.

1. Paint your nails with shimmery polish and allow to dry completely.
2. Set the white nail polish brush aside. With a toothpick, draw alternating rows of Cs lying on each nail, like this: ՍՍՍՍ ՍՍՍ
3. Allow them to dry completely.

TOTALLY LADYBUGGIN'

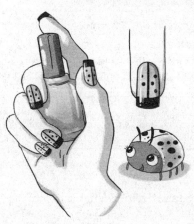

POP SOME OF THESE FASHIONABLE GOOD-LUCK CHARMS ONTO YOUR FINGERTIPS. FOR THIS PATTERN, YOU'LL NEED RED AND BLACK POLISHES AND A TOOTHPICK.

1. Paint your nails red and allow to dry completely.
2. Set the black nail polish brush aside. With a toothpick, draw a vertical black line down the center of each nail. Allow them to dry completely.
3. Use the black nail polish brush to paint the tip of each nail and allow them to dry completely.
4. Pinching the toothpick between your thumb and index finger, dot two or three spots on each half of each nail. Allow them to dry completely.

CURIOUS *Cat* QUESTIONS

CURIOSITY MAY HAVE KILLED THE CAT, BUT BEFORE THE CAT DIED, IT SURE DID SCORE A LOT OF GREAT INFO! PLUS, IT HAD EIGHT MORE LIVES LEFT, SO NO BIGGIE.

WHETHER YOU'RE WAITING FOR YOUR NAILS TO DRY, YOUR FACE MASK TO SET, OR GYM CLASS TO START, YOU CAN PASS THE TIME BY USING THESE QUESTIONS TO BECOME AN EXPERT ON YOUR FRIENDS. TAKE TURNS ANSWERING, AND WHEN YOU'VE ANSWERED ALL OF THESE, COME UP WITH SOME CURIOUS QUESTIONS OF YOUR OWN!

1. What is the most fun vacation you've been on?
2. If you were stranded on a desert island, what five foods would you want with you?
3. Would you rather jump on a trampoline made of fire or walk on a balance beam made of ice?
4. If you could wake up tomorrow as any famous person, who would you want it to be?
5. What time and place in history do you wish you could live in?
6. Chocolate or vanilla?
7. Would you rather be twenty feet tall or three inches small?
8. How many pockets is the right amount?
9. If you could have any superpower, what would you want it to be?
10. You're in outer space and you meet an alien. What is it like?
11. What animal is most like you?
12. Would you rather be able to talk with animals or read people's minds?
13. Salty or sweet?
14. What are your top three favorite movies?
15. What is your dream pet?

WHAT'S YOUR ANIMAL COMMUNICATION *Style?*

ELEPHANTS SAY "WHAT'S UP?" TO EACH OTHER BY STOMPING THEIR FEET, SEA OTTERS EXPRESS THEMSELVES BY TOUCHING NOSES, AND ROBINS SHOW THEY'RE UPSET BY ANGRY-TWEETING. (THIS IS ALSO HOW SOME HUMANS SHOW THEY'RE UPSET.)

Take this quiz to find out your own communication style, and compare your and your friends' results! If you have different styles, try doing things each other's way sometimes. That way everyllama gets a chance to communicate the way they like best!

1. YOUR FRIEND HAS A HARD DAY. HOW DO YOU MAKE THEM FEEL BETTER?
   a. Buy your friend their favorite snack.
   b. Challenge them to a cartwheeling contest.
   c. Let them vent about what happened and give advice if they ask for it.
   d. Give them one of your patented A+ shoulder massages.

2. YOU'RE HAVING YOUR FRIENDS OVER FOR A PARTY. WHICH ONE SOUNDS LIKE THE MOST FUN?
   a. Secret Santa in July.
   b. Karaoke night.
   c. Truth-or-dare extravaganza.
   d. At-home hair salon where you style each other's hair.

3. WHAT PRESENT WOULD YOU MOST WANT TO RECEIVE FROM YOUR BESTIE?
   a. A super-cute stuffed animal you've had your eye on.
   b. Tickets to a new movie you've been dying to see together.
   c. A note listing all the things your friend loves about you and your friendship.
   d. A trip to the salon for mani-pedis.

4. YOU AND YOUR FRIEND HAVE A SILLY FIGHT. HOW ARE YOU MOST LIKELY TO REACT, AND HOW DO YOU MAKE UP?
   a. Demand they give back the friendship bracelet you made, but then return it with a mega apology when you cool down.
   b. Cancel the sleepover the two of you were going to have this weekend, but then apologize and reschedule the sleepover, plus a trip to the mall.
   c. Argue your point until you've said everything you have to say . . . then, once you've gotten it all out, listen to your pal's side and come to a compromise.
   d. Give them the cold shoulder, but then make up with an apology and a big hug.

5. WHAT DO YOU LIKE TO DO AT THE PARK?
   a. Look for four-leaf clovers.
   b. Play Frisbee, catch, or tag.
   c. Gab over a picnic.
   d. Have a piggyback-ride race.

6. YOU START A YOUTUBE CHANNEL. WHAT IS IT ABOUT?
   a. Baking tasty treats.
   b. Reviewing board games.
   c. Giving friendship advice.
   d. Making up your own dance routines.

If you got mostly As, you communicate by giving presents, like a *house cat*! (Of course, a house cat thinks dead mice make good presents, so you're not exactly like a house cat.) Thoughtful presents let another person know they're on your mind.

If you got mostly Bs, you communicate by spending time with someone, like a *golden retriever*! Just like those friendly pups, you love to hang out and have fun with your best pals. This lets them know you enjoy being around them.

If you got mostly Cs, you communicate by talking about things, like a *parrot*! Chatty bird that you are, you like to tell people how you feel by using words. This helps people know exactly what you're thinking.

If you got mostly Ds, you communicate with touch, like a *bat*! Bats groom each other to get clean, but also to relax and bond, and you do the same thing. This shows people you care about them. Just be sure to ask permission before touching someone, and respect their wishes if they'd prefer a high five or a wave.

TAKE IT *Slow*

A GREAT WAY TO REV UP FOR A FABULOUS HANG-OUT SESSION WITH YOUR BEST PALS IS TO **sloooooow** DOWN. KEEPING YOUR FOCUS IN THE MOMENT, DOING SOME DEEP BREATHING, AND RELAXING YOUR MIND WILL ALL GIVE YOU THE LAID-BACK VIBES OF THE CHILLEST SLOTH IN THE FOREST. AND THEN YOU'RE READY FOR ANYTHING FROM A POP QUIZ TO A

dance party!

SLOTHFULNESS PRACTICE

THE FIRST THING TO DO FOR MAXIMUM RELAXATION IS TO KEEP YOUR MIND IN THE MOMENT. PICTURE A SLOTH—LET'S CALL HER PEGGY—HANGING FROM A BRANCH. SHE'S NOT THINKING ABOUT THE SWEET POTATO STAINS SHE GOT ON HER BEST SLOTH OVERALLS YESTERDAY OR WORRYING ABOUT WHETHER SHE'LL SCORE ANY POINTS PLAYING SLOTH-BALL THIS WEEKEND. SHE'S JUST ENJOYING THE BREEZE. SO HERE'S HOW TO BE LIKE PEGGY!

1. Close your eyes and picture the cutest, most languid sloth you can imagine. You can give your sloth an outfit, purple toes, or sequined fur. Or you can go with that old standby: Original Sloth Flavor.
2. Pick a simple task, like putting all your clothes away or drawing a picture of a toucan drinking a cup of tea.
3. Think only about what you're doing right now. *I'm picking up my jeans. I'm folding them in half, then in half again. I'm putting them in the drawer.*
4. If your mind starts to wander—say, you start imagining what snack you'll have when you're finished—close your eyes and picture your sloth standing next to you. Let your sloth take your hand and gently lead you back to your task.

LET YOUR SLOTH HELP YOU OUT WHENEVER YOU NEED TO RELAX, CALM DOWN, OR HELP TIME MOVE FASTER *OR* SLOWER. *It's always hanging around.*

HIPPO COUNT BREATHING

SOMETHING ELSE YOU CAN DO TO RELAX IS SOME DEEP, HIPPOPOTAMUS-SINKING-INTO-THE-NILE BREATHING.
(NOBODY LOVES HANGING OUT UNDERWATER AS MUCH AS HIPPOS DO, EXCEPT MAYBE FOR MERMAIDS AND THE LOCH NESS MONSTER.)
DEEP BREATHING HELPS YOUR HEART RATE SYNC UP WITH YOUR BREATH, AND THAT RELEASES ENDORPHINS, WHICH HELP YOU CALM DOWN.

1. Lie flat on your back and place a stuffed animal on your belly.
2. Take a deep breath all the way into the bottom of your belly. Your belly (and the stuffed animal on it) will rise. Your chest will also rise a little, but not as much.
3. As you breathe in, count in your head to four, like this: One hippopotamus, two hippopotamus, three hippopotamus, four hippopotamus. As you count, you can imagine yourself as a big cute hippo sinking underwater.
4. Imagine hanging out in the deep blue water with the crocodiles and turtles as you hold your breath in for another four counts: One hippopotamus, two hippopotamus, three hippopotamus, four hippopotamus.
5. Exhale to another count of four as you imagine your cute hippo self lifting your head out of the water, eyes first, then big cute hippo nostrils: One hippopotamus, two hippopotamus, three hippopotamus, four hippopotamus.
6. Repeat two more times, or until you feel nice and relaxed.

AFTER YOU'VE GOT THE HANG OF HIPPO COUNT BREATHING ON YOUR BACK, YOU CAN BREATHE THIS WAY SITTING, STANDING, OR EVEN UPSIDE DOWN. (OKAY, NOT UPSIDE DOWN. ONLY BATS ARE AUTHORIZED FOR UPSIDE-DOWN BREATHING.)

Pugicorn Pattern

YOU'RE A BUSY PUP WITH A LOT TO DO.

SO IF YOU HAVE ONLY A MINUTE TO RELAX, SET YOUR BREATHING BREAK ON TURBO SPEED WITH THE PUGICORN PATTERN. STARTING AT THE TIP OF THE HORN, TRACE THE PUGICORN WITH YOUR FINGER. BREATHE IN AS YOU MOVE DOWN THE HORN; OUT UNTIL YOU GET TO THE EAR; IN OVER THE EAR; BOTH OUT AND IN AROUND THE FACE; OUT OVER THE EAR; IN BACK TO THE HORN; AND OUT UP THE HORN.

Under the Sea MEDITATION JAR

THE ULTIMATE SPA FOR YOUR BRAIN, A MEDITATION JAR LETS YOU GO FULL JELLYFISH, FLOATING BEAUTIFULLY AND BEING CARRIED EFFORTLESSLY BY THE CURRENT.

Supplies

- CLEAN, EMPTY GLASS JAR
(OLD JAM JARS ARE GOOD, OR YOU CAN BUY A MASON JAR AT A CRAFT STORE.)
- WARM TAP WATER
- CLEAR GLUE
- BLUE FOOD COLORING
- GLITTER
- MINI SEASHELLS

Instructions

1. Have all your supplies set out and ready to go. You'll need to add everything to the jar while the water is still warm!

2. With an adult's help, fill the jar with warm tap water until it's a little under two-thirds full.

3. Fill the jar to just below the top of its curve with clear glue. The more glue there is, the longer it will take the glitter to settle.

4. Add blue food coloring one drop at a time. (It will get dark very quickly, so don't add too much at first.)

5. Screw on the lid very firmly and shake to mix together, then remove the lid and wait for the foam to disappear before starting on Step 6.

6. Add at least a tablespoon's worth of glitter in the color or colors of your choice, plus some silver or white, which will give it extra sparkle. You can use big or small pieces, or both. (Big pieces will settle faster than small ones.)

7. Screw on the lid and shake again, then remove it, add your seashells, and give it a final big shake.

8. Remove the lid and allow the contents to cool completely before putting it back on a final time.

TO USE YOUR MEDITATION JAR, SHAKE IT UP LIKE A SNOW GLOBE, SET IT DOWN, AND WATCH ALL THE GLITTER AND SHELLS UNTIL THEY'VE SETTLED TO THE BOTTOM COMPLETELY.

CELEBRATE or HIBERNATE?

ARE YOU AN OUTGOING EXTROVERT WHO LIKES TO SOCIALIZE ALL THE TIME, A QUIET INTROVERT WHO LIKES TO KEEP TO YOURSELF, OR SOMEWHERE IN BETWEEN? TAKE THIS QUIZ TO FIND OUT WHAT KINDS OF SOCIAL INTERACTIONS YOU PREFER!

1. HOW DO YOU FEEL ABOUT GROUP PROJECTS?
    a. I'd really rather work alone.
    b. I don't mind as long as my partners are my close friends.
    c. I like being able to split up the work and help each other.
    d. It's way more fun to do homework in a group than to do it alone.

2. WHICH BIRTHDAY PARTY SOUNDS BEST?
    a. A super-fab sleepover with just you and your best friend.
    b. You and your closest buds playing a ton of party games at your house.
    c. An extravaganza at an arcade with special party activities just for you and your friends.
    d. A wild dance party with the whole school invited.

3. WHAT'S THE BEST WAY TO SPEND A DAY AT A FAIR?
    a. Catching the amazing live shows.
    b. Visiting all the booths and exhibits.
    c. Funnel cake for everybunny!
    d. Hyping yourself up on every. single. ride.

4. HOW MUCH DO YOU KEEP IN TOUCH WITH YOUR FRIENDS OUTSIDE OF SCHOOL?
   a. Call or text a couple of people once or twice a week.
   b. Talk or text a few people every other day.
   c. Daily group texts.
   d. Twelve different text threads going at once, all the time.

5. WHAT DO YOU LIKE TO DO DURING LUNCH AT SCHOOL?
   a. Start on your homework or read a book.
   b. Play cards with your bestie.
   c. Chat with your friends.
   d. Get the whole lunchroom to chant, "Ta-ter tots! Ta-ter tots!"

6. WHAT ARE YOUR FAVORITE KINDS OF GAMES TO PLAY?
   a. Card games.
   b. Board games.
   c. Team games, like charades or trivia.
   d. Yard games, like tag or capture the flag.

7. WHAT SPORT DO YOU LIKE MOST?
   a. Running.
   b. Gymnastics.
   c. Basketball.
   d. Softball.

If you got mostly As, you're *very introverted*, like a *koala*! Koalas share their territory with their friends sometimes, but they also need lots of time to themselves.

If you got mostly Bs, you're *somewhat introverted*, like a *wolf*! Wolves are very comfortable with their packs, but they prefer smaller groups to big crowds.

If you got mostly Cs, you're *somewhat extroverted*, like an *ant*! Ants are great at working together with their friends, but they don't want things to get too disorganized or crazy.

If you got mostly Ds, you're *very extroverted*, like a *sparrow*! Sparrows are total party birds who love to bathe, preen, and sing with tons of their pals.

Party ANIMAL

BREAK OUT THE LEOPARD-PRINT DISCO BALL AND SET UP THE SNACK TABLE—IT'S

PARTY TIME!

Primo Penguin
PARTY PREP

WHETHER YOUR SHINDIG IS FOR TWO OR TWO HUNDRED, YOU WANT TO BE SURE YOUR GUESTS HAVE AS MUCH FUN AS A FAMILY OF PENGUINS SLEDDING DOWN A SHEET OF ICE.

PRE-PARTY PLANNING

- Get permission from your parents! (There might be things your parents don't notice, but a bunch of kids showing up in party hats is not one of them.)

- Send out showstopping invitations with the time, place, date, and any special instructions (like if it's a sleepover or if there will be swimming). You can do this online or by snail mail.

- Plan some fabulous decorations. You can buy these, make them with your very own flippers, or do both. (Hot tip: There are lots of video tutorials for cool paper crafts!)

- Figure out a delish menu. Check out Zack the Snack-Attacking Yak on page 55 for some ideas to get you started! If any of your guests have dietary restrictions, make sure you'll have plenty of food for them to eat.

- Do some pre-party cleaning! Put on some of your favorite songs to sing along to and help get the house nice and tidy.

- Don't forget your party clothes! Your penguin pals favor formal attire, of course, but you can wear whatever makes you feel fabulous.

A WADDLE OF HOSTING TIPS

- Greet each of your guests at the door as they arrive.
- Have a designated spot where guests can leave their things.
- To get things off to a fun start, set up some simple activities and games for guests to play right when they walk in. They can sink a shot in basketball toss or make an animal themed name tag.

ONCE EVERYZEBRA HAS ARRIVED, GATHER YOUR FRIENDS TO TRY OUT THE GAMES AND ACTIVITIES IN THIS SECTION, ALONG WITH ANY OTHER ACTIVITIES YOU WANT TO DO. THEN EAT SOME YUMMY FOOD AND JUST HANG AROUND HAVING A

GOOD TIME!

Bunny EARS

HOP INTO THE LAND OF FABULOUS ACCESSORIES WITH EASY-TO-MAKE ANIMAL EAR HEADBANDS. YOU CAN GO WITH BUNNY EARS, OR YOU CAN TRY OUT CAT, PANDA, DOG, OR LLAMA TOPPERS. YOU CAN EVEN DO FROG EARS! (THOSE WILL BE INVISIBLE, BECAUSE FROGS DON'T HAVE EARS.)

Supplies

- one plain, super-skinny headband for each partygoer
- poster board
- scissors
- craft glue
- markers, colored pencils, and/or crayons
- glitter, sequins, ribbons, feathers, beads, buttons, cotton balls, rhinestones, or any other craft supplies you'd like

Instructions

1. Fold a foot-long piece of poster board in half, shiny sides together.
2. Draw the outline of a stem with its base starting at the fold in the paper. The stem should be about a quarter of an inch wide and tall—about the width of a paper clip.

3. Using the top of the stem as the base, draw a bunny ear.
4. Cut out the bunny ear. Don't unfold it. Trace it onto another folded piece of poster board and cut that one out, too.
5. Decorate the fronts and backs of the ears however you'd like.
6. Unfold the bunny ears and wrap their stems around the headband. Drizzle a thin layer of craft glue all along the edges of each stem and ear and press them together firmly for about a minute to be sure they stick together. Set aside to dry for 2 hours.

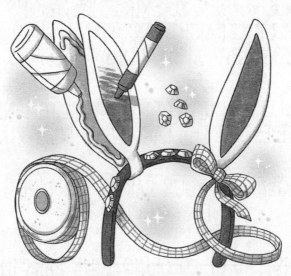

LATER, ONCE THEY'VE HAD PLENTY OF TIME TO DRY, POP YOUR EARS ON AND GO ON A HOPABOUT!

Bounce INTO SOME Decorating IDEAS!

MAKE YOUR EARS A LITTLE WIDER THAN THEY NEED TO BE, THEN GIVE THEM *rock-and-roll* FRINGE BY MAKING A BUNCH OF STRAIGHT-LINE SMALL CUTS CLOSE TOGETHER ALL AROUND THE EDGES.

USE PULLED-APART COTTON BALLS TO LINE THE "INSIDES" OF THE EARS WITH SOFT FUZZ.

GLAM UP YOUR EARS WITH LINES OF TINY RHINESTONES.

INSTEAD OF COLORING YOUR EARS IN, DRAW LOTS AND LOTS OF INDIVIDUAL LINES WITH COLORED PENCILS TO MIMIC FUR.

Flamingo Tag

FLAMINGOS LOVE HANGING OUT WITH THEIR FRIENDS

(NICE ONE, FLAMINGOS!), PLAYING ON THE LAWN (THOUGH THIS IS PRETTY MUCH EXCLUSIVE TO PLASTIC FLAMINGOS), AND STANDING AROUND ON ONE LEG FOR SOME REASON (NO ONE IS REALLY SURE WHY).

Take your flamingo vibes to new heights with a game of Flamingo Tag. It's just like regular freeze tag—but you all have to stand and run on one leg, including the person who's It. If you're tagged OR you touch the ground with your other leg, you're It.

After you've played a few rounds as flamingos, try being penguins (where you have to waddle), elephants (where you play bent over with your arm hanging down as your trunk), or any other animal you can think of. *(Inchworm Tag, anyone?)*

Animal MASH

IF YOUR WILD LIFE HAD WILDLIFE, HOW WILD WOULD YOUR LIFE BE? GRAB A PIECE OF PAPER AND SOME FRIENDS AND FIND OUT!

Instructions

1. Copy the categories on the MASH page onto your paper.

2. Player 1 closes their eyes and draws a spiral until Player 2 says stop.

3. Count the lines in the spiral—in this example there are 7.

4. Count in a clockwise motion around the page. Every time you reach 7, mark off that item. When only one item is left in a category, circle it.

5. Keep going until there's an item circled in every category—this is your full MASH report.

6. Now it's Player 2's turn, and Player 3 will say when to stop drawing the spiral.

7. Keep going till you've all had a chance to play.

8. After everybunny takes a turn, make up your own game of MASH! Take a new sheet of paper and write different selections into the categories, or create new categories altogether.

DON'T BE AFRAID TO BE *wild!*

WHAT KIND OF ANIMAL WILL YOU HAVE?

GIRAFFE

CHAMELEON

LEOPARD

OCTOPUS

HOW MANY OF YOUR ANIMAL WILL YOU HAVE?

2

5

8

11

WHAT WILL YOUR ANIMALS' FAVORITE TRICK BE?

PLAYING DEAD

PERFORMING OPEN-HEART SURGERY

TAP DANCING

LOOKING BOTH WAYS BEFORE CROSSING THE STREET

WHAT WILL YOUR ANIMALS WEAR?

TOP HATS AND TAILS

SEQUINED SHOES

TRACKSUITS

CHEF HATS AND APRONS

WHERE WILL YOU LIVE?

UNDER A ROCK

FOX DEN

CRYSTAL GROTTO

IN A HOLE IN A TREE

SO Ewe THINK EWE CAN Dance?

LAMBS AND THE MEMBERS OF THE AMERICAN BAAAAHLLET THEATER HAVE TWO THINGS IN COMMON:

1. THEY'RE GREAT DANCERS.
2. THEY HAVE EXCELLENT TASTE IN FOOTWEAR.

And ewe know what? Ewe and your flock are ALSO great dancers with excellent taste in footwear. Want to prove it? Have a dance-off—with a twist!

First, write down the names of a bunch of different animals on individual slips of paper. For maximum fun, include huge animals, teensy animals, farm animals, wild animals, animals from air, land, and sea, and any other animals you can think of.

Now place all the slips in a bowl and mix them up. Blast your favorite tunes and, one at a time, draw an animal from the bowl. Then make up a dance as that animal!

You can teach each other your dances either as you go or at the end, and you can award prizes for the wackiest, coolest, daintiest, slowest, and fastest dances.

WHAT IS YOUR *Perfect* ANIMAL LAIR?

EVERY ANIMAL HAS THEIR OWN LIVING STYLE, WHETHER IT'S UNDERGROUND, IN A TREE, OR AT THE BOTTOM OF THE OCEAN. FIND OUT WHAT YOURS IS SO YOU CAN SPRUCE YOUR ROOM UP WITH JUST THE RIGHT ANIMAL VIBES.

1. WHICH PLACE WOULD YOU MOST WANT TO LIVE IN?
   a. Log cabin surrounded by woods.
   b. Cozy cottage in a small village.
   c. House with lots of windows in an open space.
   d. Richly furnished palace.
   e. Beachside villa.
   f. Houseboat by a tropical island.

2. WHAT SOUNDS LIKE THE TASTIEST AFTERNOON SNACK?
   a. Trail mix with nuts and chocolate.
   b. Veggies and dip.
   c. Oatmeal with cinnamon and honey.
   d. Candy bar.
   e. Saltwater taffy.
   f. Chips and salsa.

3. WHAT KIND OF BOOK DO YOU MOST LIKE TO READ?
   a. Stories about witches and wizards.
   b. Anything with cute talking animals.
   c. Ones where someone goes on an adventure.
   d. Books about best friends having fun together.
   e. Stories about kids surviving in the wild.
   f. Ones about solving mysteries.

4. WHICH WOULD YOU MOST LIKE A POSTER OF?
   a. Dancing fairies.
   b. Medieval knights.
   c. Swarm of colorful butterflies.
   d. Cupcakes.
   e. Playful mermaids.
   f. Neon pineapples.

5. HOW WOULD YOU MOST WANT TO SPEND A RAINY AFTERNOON?
   a. Walking in the woods with your raincoat on.
   b. Snuggled under the covers with a book.
   c. Playing *Dance Dance Revolution*.
   d. Having a movie marathon.
   e. Swimming in the pool. (Because, hey, wet + wet = wet!)
   f. Spreading out your beach towel and building LEGO sand castles for an indoor beach day.

If you got mostly As, your perfect animal lair is in a *tree*, like an owl's! Turn your room into a real hoot by making poster frames out of twigs and branches, painting pinecones and hanging them in clusters, and covering leaves in Mod Podge and displaying them on the walls.

If you got mostly Bs, your perfect animal lair is a *burrow*, like a rabbit's! For a cozy, snug place to rest your cottontail, decorate with plenty of blankets and cushions, set up a special reading nook with a rug, comfy chair, and bookshelf, and hang scarves from the walls and ceiling.

If you got mostly Cs, your perfect animal lair is an **open plain**, like a wild horse's! To give yourself an airy space to kick up your hooves, decorate in light colors, like white, light blue, and light green. And to really transport yourself to a plain, make or buy garlands of flowers or grass to line your baseboards.

If you got mostly Ds, your perfect animal lair is a **castle**, like the queen of England's corgis'! Hey, the outdoors aren't for everypuppy. Live the life of a pampered pup by filling your room with ruffles—and you can turn simple things posh by attaching handmade fringe or tassels to them.

If you got mostly Es, your perfect animal lair is an **ocean grotto**, like a narwhal's! Give your room an underwater makeover by making construction paper jellyfish with ribbon tentacles to hang from the ceiling, taping blue tissue paper waves to the walls, and making poster frames out of seashells.

If you got mostly Fs, your perfect animal lair is in the **jungle**, like a leopard's! To turn your room green, paint huge tropical flowers on poster board and hang them on the walls, and give your storage a wild twist by using glue to cover sturdy shoe boxes with leopard print paper or fabric.

ZACK THE Snack-Attacking YAK

THREE OUT OF FOUR DENTISTS AGREE THAT THE NUMBER ONE WAY TO PREVENT A SNACK ATTACK IS TO ATTACK THE SNACKS BEFORE THEY CAN ATTACK YOU. THIS IS NOT, HOWEVER, THE CORRECT WAY TO PREVENT A BEAR ATTACK. (THE CORRECT WAY TO PREVENT A BEAR ATTACK IS TO AVOID BEARS.)

TOP SNACKING TIPS:

- Always get permission from an adult before using anything in the kitchen.
- Check to make sure you have all the ingredients on hand before you get started.
- Do an allergy check—if you or any of your friends are allergic to any of the ingredients, substitute something 100 percent less deadly.
- Clean up after yourself when you're done.

PEANUT Bunny CRACKERS

BUNNIES AND YAKS BOTH LOVE EATING GRASS AND FLOWERS, BUT ONLY BUNNIES WIGGLE THEIR NOSES WHILE DOING SO. IN HONOR OF WIGGLY NOSES EVERYWHERE, TRY OUT THESE TASTY BUNNY CRACKERS!

Ingredients

- round crackers
- peanut butter
- honey
- cinnamon
- thin apple slices
- mini marshmallows

Instructions

1. Spread a thick layer of peanut butter onto a cracker.
2. Drizzle honey over the peanut butter and top it with a sprinkle of cinnamon.
3. Add ears by pressing the tips of two apple slices into the peanut butter onto the upper right side of the cracker.
4. Add a tail by pressing a marshmallow into the peanut butter onto the bottom left edge of the cracker.

PEPPERanimal PIZZA
TINY PIZZA, BIG FUN!

Ingredients

- bagel chips
- marinara sauce
- shredded mozzarella cheese
- pepperoni and/or mushroom slices, cut in half
- ground beef and/or chopped olives
- bell peppers cut in half-length strips

Instructions

1. Spread a thin layer of marinara sauce on each bagel chip.
2. Sprinkle a thin layer of cheese on top.
3. Decorate each pizza like any animal you want! Try the pig, lion, or mouse pizzas shown here, or design your own animal.

WITH AN ADULT'S HELP, HEAT YOUR MINI PIZZAS IN THE MICROWAVE FOR ABOUT 10 SECONDS ON MEDIUM HIGH.
THEN DIG IN!

Frozen Yogurt
FRUIT BUTTERFLIES

Have you ever had butterflies in your stomach before? Well, this won't be anything like that.

Supplies

- muffin tin
- paper cup liners

Ingredients

- your favorite flavor of yogurt
- different kinds of fruit (berries and kiwi are good options)
- almonds or raisins
- chocolate sprinkles

Instructions

1. Spoon about 1/4 cup of yogurt into each cup.
2. In each cup, very carefully arrange pieces of whole or cut fruit into the shapes of two wings, with an almond or raisin for the body. Place the pieces gently on top of the yogurt and don't press them in too far.
3. Use chocolate sprinkles to make the antennae.
4. Place in the freezer for 2 hours.
5. And voilà—your plain old yogurt has metamorphosed into beautiful frozen butterflies!

CHOCOLATE *Spiders*

SPIDERS ARE GREAT AT EATING MOSQUITOES, ROLLER-SKATING, AND KNITTING. THEY'RE ALSO GREAT AT BEING A YAK-ATTACKED SNACK—WELL, WHEN THEY'RE MADE OF CHOCOLATE, ANYWAY.

Ingredients

- milk chocolate chips
- pretzel sticks
- almonds (can substitute raisins if you prefer)

Instructions

1. Prep a dipping area: line a cookie sheet with wax paper, set the pretzel sticks on one plate and the almonds on another, and leave room for the bowl of chocolate.
2. Pour one cup of milk chocolate chips into a microwavable bowl.
3. With an adult's help, heat the chips in the microwave for 20 seconds. Stir thoroughly even if the chips don't look melted.
4. Heat for another 10 seconds and stir. If needed, repeat until the chocolate is completely melted.

5. Dip the pretzels completely into the bowl of chocolate and lay 8 of them out on the cookie sheet, like this:

6. Spoon a little chocolate onto the central point where all the legs meet, then dip an almond completely into the bowl and place it on top of that.
7. Repeat until you run out of chocolate. If the chocolate begins to harden while you're still working, heat it back up (still for just 10 seconds at a time).
8. When all the spiders are assembled, stick the cookie sheet in the refrigerator for one hour. Then peel them off the wax paper and eat!

CHOCOLATE BURNS EASILY, WHICH IS WHY IT'S SUPER IMPORTANT TO HEAT IT A LITTLE BIT AT A TIME!

ANIMAL CRACKER KABLOOEY

THERE'S NO CRACKER LIKE AN ANIMAL CRACKER, 'CAUSE AN ANIMAL CRACKER IS AN EXTREMELY VERSATILE COOKIE THAT HAS BEEN POPULAR SINCE THE 1800S. BUT WHY EAT ANIMAL CRACKERS PLAIN WHEN YOU COULD HAVE ANIMAL CRACKER KABLOOEY?

Ingredients

- animal crackers, *obviously*
- your choice of
  A: frosting; peanut butter; Nutella; maple syrup; honey; or anything else that sounds delish
  B: sprinkles; M&Ms; chocolate chips; gummy bears; chopped fruit; chopped nuts; or anything else that sounds *délice* (French for "delish")

Instructions

Spread A on each cracker. Top it with B (or several Bs!). Give your animals fur or faces or fabulous outfits. Follow your heart! Or, if you're decorating a lion, your tail!

Fanciness Tip:

FOR EXTRA-COLORFUL CRACKERS, ADD FOOD COLORING TO WHITE FROSTING.

WHAT ANIMAL HAS YOUR DREAM JOB?

WHAT DO YOU WANT TO BE WHEN YOU GROW UP, AND MORE IMPORTANTLY, WHAT ANIMAL DOES THAT MAKE YOU? TAKE THIS QUIZ TO GET SOME CAREER GUIDANCE!

1. YOU WANT TO MAKE PUTTING LAUNDRY IN THE HAMPER MORE FUN. WHAT DO YOU DO?
    a. Decorate your hamper.
    b. Give yourself a star for every day you use the hamper, and when you get ten stars, reward yourself with a cool prize.
    c. Give your hamper a name and talk to it.
    d. Turn your hamper into a basketball net, complete with backboard.

2. WHICH SCHOOL SUBJECT DO YOU LIKE THE MOST?
   a. Electives.
   b. Social studies.
   c. English.
   d. Math.

3. WHICH ACTIVITY SOUNDS LIKE THE MOST FUN?
   a. Doing a photo shoot.
   b. Learning a new language.
   c. Telling a story.
   d. Experimenting with crystals by growing rock candy.

4. HOW WOULD YOU SPEND AN AFTERNOON IN THE WOODS?
   a. Building a fort.
   b. Making a map of everywhere you explore.
   c. Pretending to be a wizard.
   d. Collecting cool-looking leaves.

5. WHAT MYTHICAL CREATURE IS YOUR FAVORITE?
   a. Elf.
   b. Centaur.
   c. Dragon.
   d. Sphinx.

6. PICK SOMETHING FUN TO DO IN THE KITCHEN.
   a. Decorating cupcakes.
   b. Whisking eggs.
   c. Adding toppings to a pizza.
   d. Making a carrot stick tower for the veggie tray.

7. WHICH ANIMAL FACT IS THE COOLEST?
   a. Shrimps' hearts are in their heads.
   b. Unlike every other animal, elephants can't jump.
   c. Giraffes don't have vocal cords.
   d. Slugs have not one, not two, but four noses.

If you got mostly As, you're a *honeybee*, and you should have a creative career! Honeybees make their beautifully intricate honeycomb hives and produce lots of delicious honey. A carpenter, artist, composer, or chef also makes beautiful, useful things.

If you got mostly Bs, you're a *meerkat*, and you should have a teaching career! Meerkats teach their young how to hunt and take care of themselves. There are lots of ways for humans to be teachers—you can teach in a school, or you can teach special skills, like how to play the drums or crochet. You can even be a coach and teach people how to be better athletes.

If you got mostly Cs, you're a *dolphin*, and you should have a career that uses lots of communicating! Dolphins are super-advanced communicators—they even speak a distinct language to each other. You can be a sports writer, a playwright, or a lawyer, or you can write jokes, give speeches, or be a news anchor. The ocean's the limit!

If you got mostly Ds, you're a *beaver*, and you should work in STEM (science, technology, engineering, and math). Beavers are engineering geniuses who work together to build enormous dams. You can work as a gardener, chemist, doctor, or computer engineer to build things with other people. You can also be a zookeeper!

Horsing Around

EVERYHORSEY KNOWS THAT BEING SERIOUS IS SO FIVE MINUTES AGO—NOW IS THE TIME FOR HORSEPLAY. SO GIVE YOUR FRIENDS A GIGGLE WITH A HAYLOAD OF LAUGHS!

WHAT'S THE MATTER WITH THE SHARK'S PIANO?

IT'S OUT OF TUNE-A.

KNOCK KNOCK.
WHO'S THERE?
OWL.
OWL WHO?
OWL NEVER TELL.

WHAT WAS THE BIRD'S FAVORITE DANCE MOVE?

THE WORM.

HOW MANY CENTIPEDES DOES IT TAKE TO CHANGE FIFTY LIGHT BULBS?

ONE.

WHAT KIND OF CAT TURNS INTO A BUTTERFLY?

A CATERPILLAR.

WHAT DO YOU CALL A SINGLE EAR OF MAIZE?

A UNICORN.

WHAT DO SURFERS CALL A GIANT GNARLY WAVE?

A GNARWHAL.

WHAT'S THE LOBSTER'S FAVORITE ARCADE GAME?

THE CLAW.

WHAT SEA CREATURE IS YELLOW, ROUND, AND EATS DOTS?

PAC-MANATEE.

WHAT KIND OF CAT NEVER GETS AN ALLOWANCE?

A CHEETAH. BECAUSE CHEETAHS NEVER PROSPER.

WHAT IS THE BUNNY'S FAVORITE MOVIE?

MARY HOPPINS

WHAT INSTRUMENT DOES THE SHEEP PLAY?

TUBAAAH.

WHAT'S THE ELEPHANT'S FAVORITE PART OF A CAR?

THE TRUNK.

WHY DID THE CHICKEN CROSS THE ROAD?

TO BREAK IN HER NEW HIKING BOOTS.

WHAT DO YOU GET WHEN YOU CROSS A BUNCH OF MICE AND SOME POPSICLES?

MICICLES.

WHAT'S BLACK AND WHITE AND GRAY AND BLACK AND WHITE AND GRAY?

EVERYTHING A DOG SEES.

WHAT DOES THE TROUT LIKE TO SING?

SCALES.

WHAT DID THE CLAM RIDE AT THE UNDERSEA RODEO?

A SEAHORSE.

WHAT MEDICINE DOES A PENGUIN TAKE TO COOL DOWN?

A CHILL PILL.

THE *Birdboard* TOP TEN *Mewsic* SINGLES

1. BEARIANA GRANDE, "THANK U, HIBERNATE"
2. LUIS FOXY, "DESPAWCITO"
3. TAILOR SWIFT, "YOU BELONG WITH MEOW"
4. ED SHEEPAN, "SHAPE OF EWE"
5. PANIC! AT THE ZOO, "HI, HIPPOS"
6. CAMELA CABELLO, "CRYING IN THE DESERT"
7. BEEYONCÉ, "HONEY UP"
8. ONE DOGRECTION, "BEST BARK EVER"
9. JUSTIN BEAVER, "ONE LESS LONELY TOOTH"
10. SELLAMA GOMEZ, "THE HAIR WANTS WHAT IT WANTS"

TEN THINGS A *Snake* COULD DRESS AS FOR A COSTUME PARTY

1. JUMP ROPE
2. WALKING STICK
3. CANDY CANE
4. HULA HOOP
5. GARLAND OF FLOWERS
6. MAGIC WAND
7. FEATHER BOA
8. MICROPHONE
9. STRING OF LIGHTS
10. BEEHIVE

TOP TEN THINGS PANDAS DO WHEN THEY GO TO THE MOVIES

1. PRETEND TO BE USHERS.
2. SNEAK IN THEIR OWN BAMBOO SHOOTS TO EAT.
3. TAKE A LONG NAP.
4. CLIMB THE MOVIE SCREEN.
5. FALL OFF THE MOVIE SCREEN.
6. GET UP SEVEN DIFFERENT TIMES TO GO TO THE BATHROOM, SAYING, "EXCUSE ME, EXCUSE ME, EXCUSE ME . . ."
7. BLOCK THE VIEW OF THE PERSON SITTING BEHIND THEM.
8. HAVE A SHADOW-PUPPET SHOW IN FRONT OF THE PROJECTOR.
9. PERFORM A GYMNASTIC ROUTINE DOWN THE AISLE.
10. CLIMB INTO THE LAP OF THE PERSON NEXT TO THEM TO SNUGGLE.

TOP TEN REASONS not TO BRING A CROCODILE TO SCHOOL

1. DOESN'T FIT THROUGH THE DOOR VERY WELL.
2. ATE MY HOMEWORK.
3. TOOK A NAP ON THE TEACHER'S DESK.
4. TAIL GETS CAUGHT UNDER THE CHAIRS.
5. ATE MY LUNCH.
6. GETS ANGRY IF YOU TRY TO RIDE IT.
7. RUNS DOWN THE HALLS.
8. GROWLS LOUDLY DURING CLASS.
9. ATE OUR SCIENCE EXPERIMENTS.
10. JUMPED ONSTAGE DURING THE TALENT SHOW. REFUSED TO LEAVE.

Lightning-Round Quiz: WHAT'S YOUR UNI-PERSONALITY?

THINK QUICK! ARE YOU A LLAMACORN, PUGICORN, KITTENCORN, OR UNICORN? DON'T SPEND LONG ON THIS QUIZ—JUST GO WITH YOUR ANIMAL INSTINCTS!

1. PICK A COLOR.
    a. purple
    b. orange
    c. teal
    d. pink

2. WHAT KIND OF MUSIC DO YOU LIKE?
    a. rock/pop
    b. country
    c. hip-hop/R&B
    d. classical/jazz

3. FAVORITE JOLLY RANCHER FLAVOR?
    a. grape
    b. lime
    c. orange
    d. cherry

4. WHAT'S YOUR FAVORITE SEASON?
    a. winter
    b. spring
    c. summer
    d. fall

5. BEST PIZZA TOPPING?
    a. pepperoni
    b. mushroom
    c. peppers
    d. pineapple

6. PICK AN ANIMAL.
    a. tiger
    b. hedgehog
    c. penguin
    d. bat

7. FAVORITE FLOWER?
    a. tulip
    b. rose
    c. daisy
    d. lily

8. SELECT A SPICE.
   a. cinnamon
   b. chili powder
   c. garlic
   d. ginger

9. PICK A FAVORITE SODA FLAVOR.
   a. root beer
   b. cola
   c. lemon-lime
   d. orange

10. WHAT HAIRSTYLE DO YOU LIKE BEST?
   a. ponytail
   b. bun
   c. braids
   d. down

11. PICK A FAVORITE ANIMAL PRINT.
   a. zebra
   b. giraffe
   c. leopard
   d. snake

12. PICK A DANCE.
 a. ballet
 b. salsa
 c. hip-hop
 d. tap

13. PICK A BREAKFAST.
 a. pancakes
 b. waffles
 c. french toast
 d. omelet

14. PICK A SOUND EFFECT.
 a. Kapow!
 b. Sploosh!
 c. Biff!
 d. Vronk!

15. GET ADVICE FROM A TALKING . . .
 a. hairbrush
 b. frog
 c. mirror
 d. banana peel

16. PICK A FRUIT.
a. apple
b. strawberry
c. kiwi
d. peach

17. PICK A VEGETABLE.
a. bell pepper
b. celery
c. broccoli
d. carrot

18. SELECT A CRAFT SUPPLY.
a. sequins
b. ribbon
c. patches
d. beads

19. WHICH INSTRUMENT DO YOU LIKE?
a. drums
b. trumpet
c. piano
d. guitar

If you got mostly As, you're a *kittencorn*! You are soft and cuddly, but also sharp and pointy!

If you got mostly Bs, you're a *llamacorn*! Epic hair, full heart, can't lose!

If you got mostly Cs, you're a *pugicorn*! You are smart and loyal and run around being cute!

If you got mostly Ds, you're a *unicorn*! You are very magical, but you also like to frolic!